EXPLORING THE ELEMENTS

Aluminum

Anita Louise McCormick

Enslow Publishing
101 W. 23rd Street
Suite 240
New York, NY 10011
USA

enslow.com

Published in 2019 by Enslow Publishing, LLC.
101 W. 23rd Street, Suite 240, New York, NY 10011

Copyright © 2019 by Enslow Publishing, LLC.
All rights reserved

No part of this book may be reproduced by any means without the written permission of the publisher.

Library of Congress Cataloging-in-Publication Data
Names: McCormick, Anita Louise, author.
Title: Aluminum / Anita Louise McCormick.
Description: New York, NY : Enslow Publishing, 2019. | Series: Exploring the elements | Audience: Grades 5-8. | Includes bibliographical references and index.
Identifiers: LCCN 2017045229| ISBN 9780766099029 (library bound) | ISBN 9780766099036 (pbk.)
Subjects: LCSH: Aluminum—Properties—Juvenile literature. | Group 13 elements—Juvenile literature. | Chemical elements—Juvenile literature.
Classification: LCC QD181.A4 M33 2019 | DDC 546/.673—dc23
LC record available at https://lccn.loc.gov/2017045229

Printed in the United States of America

To Our Readers: We have done our best to make sure all website addresses in this book were active and appropriate when we went to press. However, the author and the publisher have no control over and assume no liability for the material available on those websites or on any websites they may link to. Any comments or suggestions can be sent by email to customerservice@enslow.com.

Portions of this book appeared in *Aluminum* by Heather Hasan.

Photo Credits: Cover, p. 1 (chemical element symbols) Jason Winter/Shutterstock.com; cover, p. 1 (cans) SergZSV.ZP/Shutterstock.com; p. 5 magnetix/Shutterstock.com; p. 7 Boligolov Andrew/Shutterstock.com; p. 8 NatashaPhoto/Shutterstock.com; p. 10 MikeDotta/Shutterstock.com; p. 14 concept w/Shutterstock.com; p. 20 Monty Rakusen/Cultura/Getty Images; p. 23 cigdem/Shutterstock.com; p. 24 Andrew Aitchison/In Pictures/Getty Images; p. 29 Akimov Igor/Shutterstock.com; p. 30 GIPhotoStock/Science Source; p. 33 electra/Shutterstock.com; p. 36 Patila/Shutterstock.com; p. 39 Auhustsinovich/Shutterstock.com; p. 40 Mark Williamson/Science Faction/Getty Images; p. 41 © iStockphoto.com/BanksPhotos.

Contents

Introduction 4
Chapter 1: The Story and Science of Aluminum 7
Chapter 2: Aluminum as an Element 12
Chapter 3: Understanding the Physical and Chemical Properties of Aluminum 18
Chapter 4: Aluminum Bonds to Form Compounds 27
Chapter 5: Processing Aluminum for a Worldwide Market 32
Chapter 6: Aluminum and the World We Live In 38

Glossary 43
Further Reading 44
Bibliography 45
Index 47

Introduction

Aluminum is a very important metal in today's world. Every day, nearly all of us use aluminum in one way or another. Food and drinks are packaged in aluminum cans. We use aluminum foil to wrap our leftovers. Aluminum is used in many high-tech devices. Aluminum is also used in automobile construction, especially in electric and hybrid cars that need to be as light as possible to have high mileage. Aluminum is used in many household products, as well as in rocket ships, airplanes, and railroad cars.

While aluminum can be used alone, it can also be made into compounds that combine the qualities of aluminum with that of other elements. These compounds give aluminum even more practical uses, as they combine the characteristics of more than one element.

The scientific symbol for aluminum is Al. Aluminum is a very lightweight metal. It is a silvery bluish-white color. Aluminum has the advantage of not rusting or corroding. It is strong and pliable, and it

Aluminum is the most abundant metal in the earth's crust.

can easily be formed into different shapes. It can be rolled into very thin sheets. Aluminum is also a very easy metal to recycle, which saves the cost of mining and refining it.

Many metals that are widely used in the industrial world today have been in use for thousands of years. While there is some speculation that aluminum was discovered by Romans up to 2,000 years ago, no one knew how to process it so it could be put into practical use until the 1800s. That is in part because aluminum is never found in its pure state in nature and must be extracted through special methods that were not necessary to refine other metals. But ever since scientists learned how to extract aluminum, there has been no end to the uses industry found for aluminum and the compounds that could be made with it. Many of these uses replace other metals in ways that make products lighter and less likely to corrode.

One advantage aluminum has is that it is one of the lightest of all metals. It weighs about one-third as much as an equal amount of steel. Aluminum weighs about 2.7 grams/centimeter^3, while steel weighs about 7.8 g/cm^3. This explains why aluminum has replaced steel for many uses. For example, some parts in airplanes, automobiles, and buses are now made of aluminum rather than steel because lighter vehicles use less fuel. This is very important in an age when gasoline, diesel, and jet fuel are expensive and there is a growing awareness of the negative impact burning fossil fuels has on the environment. Electric and solar-powered vehicles that are designed with a high percentage of aluminum perform much more efficiently than those made of heaver metals.

Today, aluminum and its compounds are being used in many ways that scientists of past generations could not have imagined. One of the most recent developments in the aluminum industry is a transparent compound called aluminum oxynitride, also known as ALON. It is composed of aluminum, oxygen, and nitrogen. ALON is very strong compared to glass, and it is being used in the defense industry to make bulletproof windows.

1

The Story and Science of Aluminum

Aluminum is the most abundant metal in the earth's crust. It is also the third most abundant element in the earth's crust (after oxygen [O] and silicon [Si]), making up about 8.2 percent of it. Even though aluminum is a very plentiful element, it is never found free in nature. Instead, it is always found combined with other elements.

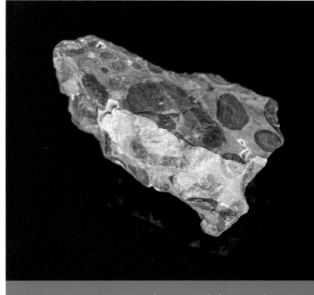

Bauxite is the world's main source of aluminum.

The most important source of aluminum is a kind of rock called bauxite (pronounced "box-ite"), which provides us with over 99 percent of the aluminum we use. Bauxite is a name for a mixture of minerals (gibbsite, diaspore, and boehmite) that contain aluminum, oxygen, and water. Bauxite is mainly located in tropical and subtropical climates. The largest amount of bauxite is found in Australia, but substantial amounts are also found in Brazil, Guinea, and Jamaica.

Early Uses of Aluminum

The word "aluminum" is derived from the Latin word *alumen*. Alum is a white powder that contains aluminum and other elements, such as sulfur (S). Ancient Roman surgeons applied alum to the wounds

Alum crystals look nothing like aluminum cans!

of injured soldiers. This closed the open blood vessels and allowed wounds to heal. Alum was also used to make dyes stick to fabric. In the Middle Ages, alum was used to cure, or dry, the skin of dead animals.

Bauxite is a mixture of hydrated aluminum oxides usually containing oxides of iron and silicon in varying quantities. Its color ranges from white or tan to a deep brown or red. The color of the bauxite depends on its components. Bauxite is soft and claylike or earthy in texture. It is characterized by having small, round, pea-sized lumps. Bauxite can easily be purified and converted into metallic aluminum.

New Discoveries in Aluminum Processing

As early as 1787, scientists began to suspect an unknown metal existed in alum. In the early nineteenth century, a British chemist named Sir Humphry Davy named the unknown metal aluminum. However, scientists did not have a way to extract this metal until 1825. The first scientist to successfully produce tiny amounts of aluminum was a Danish chemist named Hans Christian Ørsted. Two years later, a German chemist by the name of Friedrich Wohler developed a more efficient way to obtain the metal. By 1845, he was able to produce large enough amounts of aluminum to determine some of its basic properties.

In 1854, a French chemist named Henri-Etienne Sainte-Claire Deville improved on Wohler's method. Deville's method for extracting aluminum from bauxite made way for the commercial production

Aluminum at the 1855 Paris Exposition

Following French chemist Henri-Etienne Sainte-Claire Deville's development of a commercial method to extract aluminum, the metal was displayed in Paris's Universal Exposition in 1855. At the exposition, a bar of aluminum was exhibited next to the English crown jewels. Aluminum was considered a scientific marvel and was, at that time, more valuable than gold (Au).

Aluminum's display in the Paris Exhibition sparked the imagination of jewelry makers, silversmiths, and watchmakers. From 1850 to the late 1870s, aluminum made its appearance in the form of small luxury items, such as bracelets, medallions, and opera glasses.

of the metal. It also lowered the price of aluminum from $544 per pound ($1,200 per kilogram) in 1852 to $18 per pound ($40 per kilogram) in 1859.

In 1888, American chemist Charles Martin Hall, along with financier Alfred E. Hunt, founded the Pittsburg Reduction Company, which is now known as the Aluminum Company of America (ALCOA). Today, ALCOA is the world's leading producer of aluminum. ALCOA materials can be found in wheels, bike frames, airplanes, motorcycles, and soda cans.

Aluminum remained too expensive for wide commercial use until 1889, when Hall and French metallurgist Paul-Louis-Toussaint Heroult patented an inexpensive method for the production of pure aluminum. By using this new process, the cost of aluminum was reduced, paving the way for the aluminum industry to become one of the largest and most important industries in the world.

2

Aluminum as an Element

Every solid, gas and liquid is made of one or more elements. Every element is unique and composed of only one kind of atom. Like all other elements, this is true of aluminum. Atoms cannot be seen by the naked eye. It would take about five hundred million atoms to form a line that is only one inch long. Atoms are made up of even smaller components called subatomic particles.

Subatomic Particles: Neutrons, Protons, and Electrons

Atoms are made of three subatomic particles: neutrons, protons, and electrons. Neutrons and protons are clustered together at the center of the atom to form a dense core called the nucleus.

Neutrons carry no electrical charge, while protons have a positive electrical charge. This gives the nucleus an overall positive electrical charge. Since aluminum has thirteen protons in its nucleus, its nucleus has a charge of +13.

Electrons are negatively charged particles that are arranged in layers, or shells, around the nucleus of an atom. The negative electrons are attracted to the positive nucleus, and it is this attraction that keeps the electrons spinning rapidly around the nucleus and holds the atom together. The number of protons and electrons in a neutral atom are equal, so the positive and negative charges of the atom are balanced. Therefore, since aluminum has thirteen protons, it also has thirteen electrons.

The Periodic Table: An Organization System for the Elements

There are more than one hundred known elements today. As scientists discovered more elements over the years, they realized the elements needed organization. Eventually, elements were arranged on a chart called the periodic table.

The periodic table we use today is based on the work of a Russian chemist named Dmitry Mendeleev. He published the first version of the periodic table in 1869, while teaching chemistry at the University of St. Petersburg in Russia. Mendeleev sought to

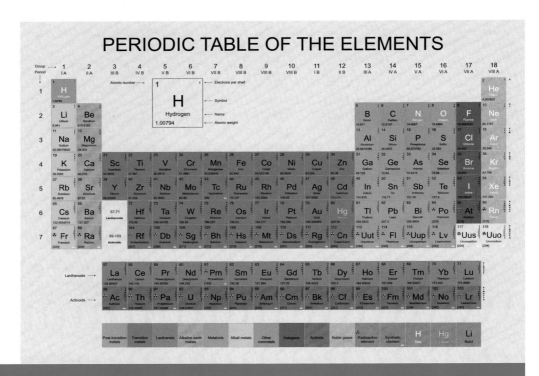

The periodic table of elements organizes elements into groups and periods.

organize the elements in a way that would make it easier for his students to study and understand them.

Mendeleev loved to play cards, so he decided to use cards as a way of organizing information about the elements. After writing information about the elements on cards, he arranged them in groups, according to their properties. He then created the periodic table, organizing the elements in horizontal rows, according to weight, with the lightest element of each row on the left end and the heaviest on the right.

Unlike Mendeleev's chart, the periodic table we use today lists the elements in order of increasing atomic number. An element's atomic number is equal to the number of protons in its nucleus. By seeing where an element is located on the periodic table, you can predict whether it is a metal, a nonmetal, or a metalloid.

Metals, such as aluminum, are easily recognized by their physical traits. Generally, metals can be polished to be made shiny. Most metals have the ability to be hammered into shapes without breaking, a property called malleability. Metals are usually ductile, meaning that they can be pulled into wires. They also conduct electricity. Aluminum has all of these physical traits.

Substances such as wood, glass, or plastic are classified as nonmetals because they lack the characteristics of metals. Metalloids, or semimetals, have characteristics of both metals and nonmetals.

If you look at the periodic table, you will notice the elements are divided by a "staircase" line. The metals are found to the left of

Properties of Aluminum

Chemical Symbol: Al
Properties: Lightweight, ductile metal that is solid at room temperature, conducts heat well, and does not corrode.
Discovered By: Although it is impossible to say when aluminum was first discovered, a method of extracting the metal was developed in 1825 by Hans Christian Ørsted.
Atomic Number: 13
Atomic Weight (actual): 26.982 atomic mass units (amu)
Protons: 13
Electrons: 13
Neutrons: 14
Density at 68°F (20°C): 2.7 g/cm^3
Melting Point: 1221°F (660°C)
Boiling Point: 1221°F (660°C)
Commonly Found: In the earth's crust

this line, the nonmetals on the right. Most of the elements bordering the line are metalloids. Though aluminum is one of the elements that border the staircase line, it is the one exception to the rule.

Every Element Is Unique

What makes aluminum different from other elements, such as oxygen or silver (Ag)? The difference lies in the number of protons that are found in the nuclei of its atoms. Aluminum, which has thirteen protons in its nucleus, has an atomic number of thirteen. On the periodic table, this number is found above the element's symbol. The fact that aluminum has thirteen protons in its nucleus is what makes it aluminum. If one proton was added to aluminum's nucleus, it would become silicon, which has fourteen protons in its nucleus. If one of aluminum's protons was removed, it would become magnesium (Mg), which has twelve protons in its nucleus.

Understanding Element Groups and Periods

As you look across the periodic table from left to right, each horizontal row of elements is called a period. Elements are arranged in periods by the number of electron shells that surround the nuclei of their atoms. Aluminum is in period 3, so each of its atoms has three shells of electrons surrounding its nucleus. There are two electrons in aluminum's innermost electron shell, eight in the next, and three in its outermost shell. The electrons in the outermost shell are called valence electrons. These electrons determine how an element acts.

As you read down the periodic table from top to bottom, each vertical column of elements is called a group. All of the elements in a group have the same number of electrons in their outermost electron shell. Aluminum is in group IIIA. The other elements found in group IIIA are boron (B), gallium (Ga), indium (In), and thallium (Tl). Each of these elements has three electrons in its outermost electron shell. However, they have very little else in common with one another and differ widely in the way that they react with other chemicals.

The group IIIA elements become more metallic in nature when moving down the column. Although they are quite different from one another, all of the group IIIA elements (except boron) have a tendency to lose three electrons to form a triply charged positive ion. An ion is a charged particle. It is formed when an atom gains or loses electrons from the shells that surround the nucleus. Atoms are usually electrically neutral, which means they carry no charge. They carry no charge because they have an equal number of positively charged protons and negatively charged electrons. However, if an atom picks up extra negatively charged electrons, it becomes a negatively charged ion (called an anion). In the same manner, if an atom loses electrons, it becomes a positively charged ion (called a cation).

3
Understanding the Physical and Chemical Properties of Aluminum

All elements have characteristic physical and chemical properties. These properties help scientists to identify and classify them. Some examples of physical properties are an element's phase at room temperature, density, and hardness. The chemical properties of an element describe the element's ability to undergo chemical change. A chemical change converts one kind of matter into a new kind of matter. If an element undergoes chemical change easily, it is said to be very reactive.

Aluminum Is Solid at Room Temperature

At room temperature, an element is found in one of three phases: solid, liquid, or gas. Knowing the phase, or physical state, of an element at room temperature helps scientists to identify it. Aluminum is found in the solid phase at room temperature. All metals are solids at room temperature, except for mercury (Hg), which is a liquid. A solid has a fixed shape and volume. Solids also resist being compressed and having their shape changed.

Aluminum Has a Melting Point

If solid aluminum is exposed to a high enough temperature, it will turn to liquid. The temperature at which this phase change occurs is aluminum's melting point. For a solid to melt, the forces holding its atoms together must be overcome. Metals often have high melting points due to the strong bonds that hold their atoms together. Even so, there is a considerable variability in the melting points of metals. Mercury, for example, melts at about -38° Fahrenheit (-39° Celsius), while tungsten (W) melts at about 6,170° F (3,410° C), the highest melting point of any metallic element. Aluminum's melting point lies somewhere in between these two elements, at 1,221° F (660° C).

Aluminum can be heated until it melts and can be molded into all kinds of shapes.

Aluminum's Density Is Light for a Metal

Density is another physical property of matter, and each element has a unique density. Density measures how compact an object is—in other words, how much mass it contains per unit of volume. Solids often have higher densities than liquids, which, in turn, have higher densities than gases. Aluminum has a density of 2.7 g/cm³.

In chemistry, the densities of many substances are compared to the density of water (H_2O), or 1.0 g/cm³. If an object with a lower density than water is placed in water, it will float. However, if an object has a higher density than water, it will sink. Because aluminum has a higher density than water, pieces of aluminum will sink when dropped in water. Metals generally have high densities. Only sodium (Na), potassium (K), and lithium (Li) have densities lower than that of water.

Though aluminum has a higher density than water, it is one of the lightest of all metals. It weighs about one-third as much as an equal amount of steel. Aluminum weighs about 2.7 g/cm^3, while steel weighs about 7.8 g/cm^3. This explains why aluminum has replaced steel for many uses. For example, some parts in airplanes, automobiles, and buses are now made of aluminum rather than steel because lighter vehicles use less fuel.

Aluminum's Hardness Affects Its Strength

Aluminum is a very soft metal. Although this makes the metal very easy to shape, it means that objects made from aluminum are not very strong. On Mohs' hardness scale, aluminum has a hardness of 2.75. For comparison, your fingernail has a hardness of 2.5, and a penny has a hardness of 3.5. Aluminum is harder than your fingernail and could scratch it. However, aluminum is softer than a penny and could, in turn, be scratched by it. However, by mixing aluminum with other metals, such as copper or magnesium, it can be made as hard as steel.

Mohs' Scale Rates Elements' Hardness

Mohs' scale is used to classify the hardness of minerals, metals, and other materials. This scale was published in 1822 by an Austrian mineralogist named Friedrich Mohs. Mohs got the idea for the scale from observing miners, who routinely performed scratch tests.

The scale shows ten levels of minerals, in order of increasing hardness. Each successive mineral is able to scratch the preceding mineral and can be scratched by all that follow it.

Examples of Hardness Rating
1. Talc
2. Gypsum (rock salt, fingernail)
3. Calcite (copper)
4. Fluorite (iron [Fe])
5. Apatite (cobalt [Co])
6. Orthoclase (rhodium [Rh], silicon, tungsten)
7. Quartz
8. Topaz (chromium [Cr], steel)
9. Corundum (sapphire)
10. Diamond

Aluminum Conducts Electricity and Heat

Like other metals, aluminum is able to conduct electricity and heat. Aluminum is a good conductor, and electrical current and heat are able to move through aluminum easily. Metals are able to conduct electricity because electrons from the outer shells of the atoms move from atom to atom in what is called a sea of electrons. As these electrons move, they carry the charge, and thus electricity, with them. Aluminum, along with copper and silver, is one of only three metals that are used to make electrical conduction wires. Though aluminum does not conduct electricity as well as copper, its light weight makes it ideal for use in things such as overhead power cables.

The free-moving electrons in metals also make them good conductors of heat. When metals such as aluminum are heated, the electrons gain more

Aluminium can be rolled into thin sheets and used for decorative purposes.

energy. This makes them move about more quickly, distributing the heat throughout the metal. Aluminum is such a good conductor of heat that it is commonly used to make saucepans. Not only is aluminum an inexpensive choice for such purposes, but it also transmits heat efficiently and cools down very quickly.

Aluminum Reflects Both Light and Heat

One of aluminum's most useful properties is its ability to reflect light and heat. Much of the light and heat that strike the surface of aluminum bounce off.

A cross section of a high-tension power line shows seven strands of steel wrapped in three layers of aluminum.

Aluminum reflects about 80 percent of the light that hits its surface. Many mirrors contain aluminum. Most mirrors are made up of three layers: a protective bottom layer, a middle layer of metal (such as aluminum, silver, or tin), and a glass top layer. Because of aluminum's reflective property, the metal is also widely used as a reflector in light fixtures.

Aluminum also reflects nearly nine-tenths of the heat that reaches it. For this reason, aluminum is often used in housing insulation and as roofing material. Aluminum can be used to direct heat inside the house, or it can be used to reflect heat away from the house to keep it cool.

Aluminum is also used to make the suits worn by firefighters. These special aluminum-coated suits reflect heat, helping to keep the firefighters safe as they walk through flames. Astronauts also wear suits coated with aluminum. These spacesuits help to regulate body temperatures of the astronauts by preventing extreme heat gain or loss.

Aluminum Foil Blocks Cell Phone Signals

Besides being able to reflect light and heat, aluminum can also block radio signals. This includes the radio signals used to operate cell phones. If you wrap a piece of aluminum foil around a cell phone, it cannot receive or make calls. This is because the aluminum foil enclosure has created what is known as a Faraday cage.

Scientists use Faraday cages to keep areas free of radio signals when this is necessary to perform certain kinds of electronic experiments or test devices in a controlled environment.

Faraday cages do not need to be made from solid metal to work. As long as there are no holes or openings larger than the length of the shortest wave you want to block, it will work.

The door of a microwave oven is one example of a Faraday cage. Even though it has holes so you can watch the food heat inside, the screen on the door blocks the microwave signals that are used to cook food from going beyond the oven.

This also explains why it can be difficult to get good phone reception in buildings that contain a lot of aluminum.

Aluminum Reacts and Forms Bonds

Aluminum easily reacts with other elements, especially oxygen. Aluminum has only three electrons in the outermost shells of its atoms; since this outermost shell can potentially hold eight electrons, aluminum is said to have an incompletely filled shell. When atoms have incompletely filled shells, they will generally either give up those electrons to other atoms, forming what is called an ionic bond, or join together with other atoms and share them, forming what is called a covalent bond.

When elements form bonds with one another, they create compounds. Because aluminum is so reactive, in nature it is always found in compounds. This is one reason it took so long for scientists to discover pure aluminum.

Aluminum Bonds to Form Compounds

A compound is formed when two or more elements are bonded together. There are millions of different compounds all around you. When elements join together to form compounds, they lose their individual traits. Though aluminum alone is very reactive, compounds containing it can be quite stable. The traits of a compound are often very different from the traits of the individual elements from which it is made.

Aluminum is very soft, but it can be made extremely hard by bonding it with elements such as copper, magnesium, or zinc (Zn). Aluminum itself is light, but it can be made even lighter if it is bonded to the lightest metal, lithium. Some important aluminum compounds are the oxides and the sulfates.

Alumina, also Known as Aluminum Oxide

Aluminum oxide (Al_2O_3), or alumina, is probably the most important compound of aluminum. When exposed to air, aluminum quickly reacts with the oxygen in the air and becomes coated with a thin film of aluminum oxide. The aluminum oxide is a transparent film that adheres very tightly to the surface of the aluminum, making it very difficult for oxygen to react with the metal beneath. For this reason, aluminum is usually considered to be corrosion-resistant. Aluminum cans do not rust the way steel cans do.

Most alumina is used to make aluminum metal, but a large amount is also used for other purposes. Alumina can form corundum, one of the hardest materials in the world. Its crystals are so hard that they are often used as an abrasive in sandpaper and for various grinding tools. Because of its high melting point, it is used to make firebricks that line the inside of ovens and furnaces. It is also used in the cosmetics industry in lotions and creams. Large crystals of corundum that contain traces of other metals are valued as gems. For example, ruby is aluminum oxide that contains a small amount of chromium. Sapphire is aluminum oxide that contains trace amounts of iron and titanium (Ti). Though crystals of

Two hundred billion aluminum cans are used worldwide every year for beverages.

pure aluminum oxide are colorless, these other trace metals give the gems various beautiful colors.

Aluminum Sulfate Has Many Practical Uses

Aluminum sulfate ($Al_2[SO_4]_3$) is a compound of aluminum that is produced in very large quantities. Aluminum sulfate is sometimes called pickle alum due to its use in giving sourness to pickles. A lot of aluminum sulfate is used in the paper industry. When making printing paper, materials such as clay and rosin (a tree resin) are added to improve the paper's ability to hold ink. Aluminum sulfate is needed in order to attach the clay and rosin to the paper fibers.

Aluminum sulfate is also used to treat wastewater. It changes the surface characteristics of suspended solids, or substances floating in the water, so that they will attach to one another. Antiperspirants also contain aluminum sulfate, which acts as an astringent, or a substance used to close the openings of sweat glands.

When aluminum oxide is hydrated (containing water) it is known as aluminum hydroxide (Al_2O_3 $3H_2O$). This compound is a white, jelly-like substance that is formed when an alkali, or base, is added to an

Aluminum can take many forms, from hard turnings to softer hydrated aluminum sulfate.

aluminum salt. Toothpaste often contains small amounts of aluminum hydroxide. It helps to counteract the buildup of acids in the mouth that destroy the enamel layer of our teeth.

Aluminum and Chlorine Bond into Aluminum Chloride

Aluminum chloride ($AlCl_3$) is a white solid. It is made in a laboratory by passing dry chlorine (Cl_2) gas or dry hydrogen chloride (HCl) gas over a heated sample of aluminum. Aluminum chloride reacts readily with water and turns to fumes when in contact with moist air. Like aluminum sulfate, aluminum chloride is an astringent and is therefore an ingredient in many antiperspirants.

Aluminum Compounds Help Put Out Fires

Aluminum sulfate, when combined with sodium bicarbonate ($NaHCO_3$), can be a source of carbon dioxide (CO_2) and foam. This combination was used for years to make fire extinguishers, the compounds kept in separate compartments of the extinguisher. When the fire extinguisher was needed, the seal separating the compartments was broken. The resulting mix was aluminum hydroxide (Al_2O_3 $3H_2O$) and carbon dioxide gas. The gas couldn't escape through the sticky liquid. It bubbled through it, creating a foam. This foam blanketed the fire with material that could not burn. This kept oxygen from feeding the flames and put out the fire.

5
Processing Aluminum for a Worldwide Market

Every year, worldwide demand for aluminum increases. Aluminum comes from two sources: primary metal, which is produced from ore, and secondary metal, which is produced from scrap. According to the Aluminum Association, nearly 75 percent of all the aluminum that has ever been produced is still being used in some way. Some of the biggest producers of aluminum in the world are China, Russia, and North America. While figures on aluminum recycling, use, and production rapidly change, industry, government, and

A truck moves bauxite at a mine near Weipa, Queensland, in Australia.

education websites listed in the back of the book are usually the best places to go for up-to-date information.

Aluminum Is Extracted from Ore

Most of the aluminum in the world today is made from bauxite. Bauxite is composed mostly of aluminum hydroxide, meaning that it is made of alumina and water. Since bauxite is found close to the surface of the ground, mining it is fairly easy. Bulldozers expose bauxite deposits by clearing away vegetation and topsoil. The bauxite is broken up with explosives and gathered into trucks and trains by powerful mechanical diggers.

After the bauxite has been mined, it is sent to processing plants where the water is removed, leaving behind alumina. Australia, the United States, and China are the world's largest producers of alumina. Because it does not contain water, alumina is much lighter than bauxite. It flows easily through the processing plants, unlike bauxite, which has a sticky, muddy consistency. The inexpensive process developed to extract the alumina from bauxite is called the

You Can Make a Fire with Aluminum

You can make a fire with an aluminum can and a bar of chocolate! The bottom of the can is ideal for focusing and reflecting the sun's light and energy.

If you look at the bottom of an aluminum can, you will notice that it is quite dull. That is where a chocolate bar comes in handy. The chocolate does an excellent job of polishing the aluminum. By rubbing chocolate on the bottom of the can and then polishing it with the wrapper or a piece of cloth, you can get it to shine. (Do not eat the chocolate afterward, however, because it will pick up aluminum from the can.)

Once the can is shiny, you need to find a suitable piece of tinder and wait for a sunny day. Aim the bottom of your can at the sun and direct the reflected light at the tinder. You will have a campfire in no time.

When you are experimenting with fire, it is important to remember the safety rules. Fires can start easily and spread quickly. Make sure to work in an area that is not too close to trees, dry bushes, grass, or leaves. Do not try to make a fire if it is a windy day. It is also good to have an adult present that can help put out the fire in case it gets out of control.

Bayer process. It is named after Austrian chemist Karl Joseph Bayer, who pioneered the process in 1888.

How the Bayer Process Works

The Bayer process refines bauxite into almost pure alumina by mixing bauxite with sodium hydroxide (NaOH) and water at a high temperature and pressure. This results in a boiling hot solution of sodium aluminate (NaAl[OH]J. This solution is drained into tanks, where impurities are filtered out. The resulting liquid is cooled in vats.

As the liquid cools, aluminum hydroxide crystals form. The crystals are washed and heat-dried in ovens at temperatures of over 1760° F (960° C) in order to drive off any remaining moisture. The resulting white, granular powder is alumina. This alumina is sent to a refinery, where pure aluminum will be produced from it.

How the Hall-Heroult Process Works

Aluminum metal is refined from alumina in a process called the Hall-Heroult process, named after Charles Martin Hall and French metallurgist Paul-Louis-Toussaint Heroult, who discovered the process independently of one another in 1886. In the Hall-Heroult process, alumina is dissolved in molten cryolite (Na_3AlF_6). Cryolite is another aluminum-containing mineral. In addition to aluminum, cryolite contains sodium and fluorine (F). At one time, cryolite was found in large quantities on the west coast of Greenland.

Scrap aluminum is piled in a yard, ready for recycling and reuse!

However, that supply ran out in 1987. Though small quantities can still be found in various locations around the world, the use of cryolite has mostly been replaced by artificially produced sodium aluminum fluoride.

Dissolving the alumina in the molten bath allows it to break down into its ions, Al^{3+} and O^{2-}. Electricity is passed through the molten bath in a process called electrolysis. Since the oxygen ions are negatively charged, they move through the solution to the positively charged electrode of the bath, or the anode. Here, free oxygen, O_2, is released.

Meanwhile, the positively charged aluminum ions are drawn toward the negatively charged electrode, or cathode. This causes a layer of molten aluminum to form at the bottom of the bath. For about every four tons of bauxite used in these processes, a ton of aluminum will be produced.

Recycling Aluminum from Scrap Saves Energy

Obtaining aluminum through recycling instead of through the refinement of aluminum ore saves a considerable amount of energy and money. Because scrap aluminum has already been refined, the energy needed for the Bayer and Hall-Heroult processes is saved when aluminum is recycled. Aluminum scrap needs only to be melted down before it can be reused. Therefore, recycling uses about one-twentieth of the energy it took to produce the aluminum in the first place. Fewer resources, such as coal and oil, are used, and fewer pollutants, such as carbon dioxide and sulfur dioxide (SO_2), are released. Recycling also eliminates the cost of mining and shipping.

6

Aluminum and the World We Live In

From aluminum cans to airplane parts to electrical wires, the contribution that aluminum has made to modern society is impossible to measure. We use aluminum and its compounds when we cook, when we travel, and even to boost rockets into space. Many electronic devices we use contain aluminum and aluminum compounds. Aluminum compounds are even found in our food, toiletries, and cosmetics.

Aluminum Compounds and Your Health

Compounds that contain aluminum are sometimes found in the human body. Though there is no evidence that aluminum is harmful in small doses, physicians believe that a buildup of the metal in the body may cause health problems. Aluminum can enter the body in many ways.

Some antacids, drugs used to reduce the amount of acid in the stomach, contain aluminum. Aluminum may also enter foods prepared with aluminum pots, utensils, or foil. Some food additives, such as potassium alum ($KAl[SO_4]_2$), used to whiten flour, also contain aluminum. As we have seen, aluminum chloride and aluminum sulfate are used to make antiperspirants, and toothpastes contain aluminum hydroxide. Other aluminum-containing compounds are also used in cosmetics.

Aluminum cooking pots may not be the best choice for healthy living.

Though it is believed most aluminum is excreted from the body, some can build up in the brain, thyroid gland, liver, and lungs. Some scientists believe that aluminum is linked to diseases such as emphysema, fibrosis, lung disease, and Alzheimer's disease. There is some evidence that Alzheimer's disease, which destroys a person's memory, is more common in areas where the water contains a lot of aluminum.

Aluminum and the Space Program

A rocket is a vehicle that carries objects through air and space. Today, rockets carry explosive devices to targets, boost satellites and other spacecraft into space, and carry scientific instruments into the upper atmosphere. Rockets are propelled by both fuel and an oxidizer, or an oxygen-containing substance.

There are two types of rocket fuel: liquid propellants and solid propellants. Liquid propellants are usually composed of liquid hydrogen and liquid oxygen. Solid rocket propellants contain aluminum metal powder and an oxidizer of ammonium perchlorate (NH_4ClO_4). Without the solid rocket boosters that use this fuel, the space shuttles would not have been able to make it out of orbit.

Aluminum has been an important part of the space program, from rockets to fuels!

In a landfill, it takes four hundred years for aluminum to deteriorate. Don't throw those cans away. Recycle!

Recycling Aluminum Helps the Environment

Though there is still plenty of aluminum in the ground to be mined, bauxite mining has already devastated large areas of land. There is no limit to the number of times aluminum can be recycled, yet every day, one hundred million beverage cans are sent to landfills, incinerated, or thrown on the ground in the United States. Since aluminum does not usually corrode or rust, aluminum cans may remain intact for decades.

Facts About Aluminum Drink Cans

- Aluminum drink cans were first manufactured in 1958.
- Globally, nearly 70 percent of aluminum cans are recycled.
- In the United States, more than 150 million cans a day are sent to recycling.
- In a landfill, it can take four hundred years for aluminum cans to deteriorate.

There are many ways that you can help. Find local recycling centers in your area. You can take your cans to a recycling center, or if your community offers curbside recycling, you can place your empty aluminum cans in the bins that are provided. One hundred percent of aluminum cans that are recycled become new aluminum products in as little as sixty days.

Glossary

atom The smallest part of an element having the chemical properties of that element.

bauxite The rock from which we get the most aluminum.

bond An attractive force that links two atoms together.

compound Two elements bonded together.

covalent Chemical bonds formed by the sharing of electrons between atoms.

crust The surface layer of the earth.

electrolysis The process in which electricity is passed through a liquid between electrodes.

Faraday cage An enclosure used to block electromagnetic signals.

mass The amount of matter an object contains.

matter Anything that takes up space and has mass.

ore A mineral deposit containing something that can be profitably mined.

oxide A compound that contains oxygen.

pure Referring to a material that contains a single kind of atom.

refine To purify something.

volume The amount of space that something occupies.

Further Reading

Books

Baby Professor. *The Periodic Table of Elements—Post Transition Metals, Metalloids and Nonmetals.* Amazon Digital Services: Baby Professor, 2017.

Callery, Sean, and Miranda Smith. *The Periodic Table.* New York, NY: Scholastic Nonfiction, 2017.

Csiszar, John. *Aluminum—Chemistry of Everyday Elements.* Broomdale, PA: Mason Crest, 2017.

DK. *The Elements Book: A Visual Encyclopedia of the Periodic Table.* New York, NY: DK Children, 2017.

Websites

Geology.com
geology.com/minerals/bauxite.shtml
Explore the main source of aluminum.

Jefferson Lab
education.jlab.org/itselemental/ele013.html
Learn historical and scientific facts about aluminum.

WebElements
www.webelements.com
Explore an interactive table of elements.

Bibliography

American Chemical Society. "Production of Aluminum: The Hall-Héroult Process." Retrieved September 6, 2017. https://www.acs.org/content/acs/en/education/whatischemistry/landmarks/aluminum-process.html.

Brady, James E., and Frederick A. Senese. *Chemistry: Matter and Its Changes. 5th ed.* New York, NY: John Wiley & Sons, 2007.

Ebbing, Darrell and Steven D. Gammon. *General Chemistry. 11th ed.* Pacific Grove, CA: Brooks Cole, 2016.

Farndon, John. *The Elements: Aluminum.* New York, NY: Marshall Cavendish Corporation, 2001.

Pappas, Stephanie. "Facts about Aluminum." Live Science, September 28, 2014. https://www.livescience.com/28865-aluminum.html.

Regan, Sean Michael. "The Wonders of Transparent Aluminum." Make. Retrieved September 12, 2017. https://makezine.com/2012/01/17/transparent-aluminum.

Stwertka, Albert. *A Guide to the Elements*. 3rd ed. New York, NY: Oxford University Press, 2012.

US Geological Survey. Mineral Commodity Summaries, January 2016. Retrieved September 6, 2017. https://minerals.usgs.gov/minerals/pubs/commodity/aluminum/mcs-2016-alumi.pdf.

West Larry, "The Benefits of Aluminum Recycling." Thoughtco, March 29, 2017. Retrieved September 12, 2017. https://www.thoughtco.com/the-benefits-of-aluminum-recycling-1204138.

Index

A
ALCOA, 11
alumina, 28, 33, 34, 35–36
aluminum
    compounds, 26, 27–31, 38
    extracting/processing, 5–6,
        9–11, 33–37
    history of use, 5, 8–9, 10
    in human body, 38–39
    properties, 4–5, 6, 7, 15, 16,
        17, 19–26, 27–28
    recycling of, 5, 32, 37, 41–42
    sources of, 8, 9, 32
    uses for, 4, 6, 11, 21, 22, 24,
        25, 28, 29–31, 38–42
aluminum chloride, 31, 39
aluminum hydroxide, 30–31, 33
aluminum oxide, 28–29
aluminum oxynitride (ALON), 6
aluminum sulfate, 29–30, 31, 39
atomic number, 15, 16
atoms, 12–13, 16, 17, 22, 26

B
bauxite, 8, 9, 33–34, 35, 37, 41
Bayer process, 35, 37

C
cans, aluminum, 4, 11, 28, 42–43
compounds, 4, 6, 26, 27–31, 38
conductivity, 22–23
corundum, 28
covalent bond, 26
cryolite, 35–36

D
Davy, Sir Humphry, 9
density, 20–21

Deville, Henri-Etienne Sainte-Claire, 9, 10

E
electrolysis, 36–37
elements
    periodic table and, 13–17
    properties of, 18–26

F
Faraday cage, 25–26
fire, starting a, 34
foil, aluminum, 4, 25, 39

H
Hall, Charles Martin, 11, 35
Hall-Heroult process, 35, 37
hardness, 21, 22
Heroult, Paul-Louis-Toussaint, 11, 35
Hunt, Alfred E., 11

I
ionic bond, 26

M
melting point, 19, 28
Mendeleev, Dmitry, 13–15
Mohs' scale, 21, 22

O
Ørsted, Hans Christian, 9

P
Paris Exhibition (1855), 10
periodic table of elements, 13–17

R
radio signals, blocking, 25–26
recycling, 5, 32, 37, 41–42
reflection, 23–26
rockets, 4, 38, 40

S
subatomic particles, 12–13, 15, 16, 17, 22–23, 26

W
Wohler, Friedrich, 9